THe green Tomato

cookbook

By Paula Simmons

Drawings by Ruth Richardson

Published by
PACIFIC SEARCH
715 Harrison Street
Seattle, Washington 98109

Cover and Page Design by Frederick Walsh

First printing, June 1975

Copyright © 1975 by Pacific Search
International Standard Book Number 0-914718-08-8
Library of Congress Catalog Card Number 75-12073
Printed in the U.S.A.

TABLE OF CONTENTS

GROWING AND COOKING HINTS

No one grows green tomatoes on purpose. But whether or not the weather is cooperative, the first fall frost is bound to leave you with more unripe tomatoes than you can store or use. If you're like me, you hate to throw away anything edible. Don't despair! You'll find in this book dozens of imaginative ways to use and preserve excess green tomatoes: breads, desserts, casseroles, vegetable dishes, pickles, relishes, and mincemeat. You'll find yourself positively luxuriating in that oversupply of a delicacy you can hardly ever buy at the supermarket. There are a few basic secrets of success involved in growing tomatoes. They need a very sunny location and good drainage. Since they are deep-rooted, the ground should be cultivated deeply before planting. Fertilizing should be light until fruit forms; otherwise you'll have more lush green leaf growth than fruit. You can use one of the blossom-setting hormone preparations if summer nights are too cool in your area, or if the days are either too hot or too rainy.

You can start tomatoes indoors from seed six to eight weeks before time for transplanting outdoors. If you prefer to buy nursery-grown plants, be sure to buy short, stocky ones. A third method is to seed right into the garden with a variety like Cold Set.

You can grow tomatoes even though you lack garden space. "Patio variety" tomatoes adapt well to container-growing on a patio, windowbox, or balcony. Containers have the advantage of outwitting slugs, cutworms, and wandering dogs and cats, but tend to dry out rapidly, so be sure to water them frequently. Wooden boxes or tubs with good drainage are better than plastic or clay pots. You can grow dwarf types in gallon-size containers and standard ones in two- or three-gallon containers provided they have adequate drainage.

Choose varieties developed specifically for your area or climate. If in doubt, raise several kinds so you can be sure of a good yield from at least one. Cherry tomatoes or small standard varieties ripen faster, so are more suitable for areas with cool summers. If you don't enjoy garden work, try varieties advertised as needing no pruning, staking, or spraying. Nematode-resistant strains are well worth the trouble of seeking out.

Here are a few interesting varieties: Caro Red (Guerney), extra high in vitamin C; White Beauty, turns white when ripe, low-acid; Tiny Tim, midget plant, small tomatoes, good either in garden or containers; Jubilee (Burpee), bright golden-orange fruit, high yield; Roma VF, bright red plum-shaped, excellent for tomato juice or paste, or canning whole; Spartan Red, sparse foliage, requires less sunlight; Climbing Tomatoes, for fence or trellis, require long season or hot weather, not recommended where summers are cool.

Early varieties: Big Early (Burpee), large; Sensation hybrid (Shumway), extra early, large-fruited, high yield; Early Girl hybrid, medium size; Earliana, dependable; Cold Set, for planting directly into garden; Springset, sets blossoms even in cold weather; Jet Star (Harris), large tomato; Fire Ball, very early, medium-small tomato; Moreton hybrid, very large tomato; Early Salad hybrid, small plant, small tomatoes, extremely early; Bonny Best (Nichols), a favorite in the north; Spring Giant F-1 hybrid (Nichols), high yield, half-pound fruit; Pixie hybrid, ripens well in Northwest. Sources given for seed are not necessarily the only source of that variety.

Seeds may be started in peat pots or egg cartons (cardboard, not plastic). Plant two seeds to a pot or egg cup, using commercial planting mix or your own soil-sand-peatmoss mixture. For ease of handling, place peat pots on the white plastic trays used to package meats. Enclose trays, or egg cartons with tops removed, in clear plastic bags, loosely tied so that the plastic does not touch the soil. Then keep them in the dark, at 70 to 80 degrees. The constant humidity and warmth will hasten germination. Enclosed in plastic, the plants probably will not need watering during the 7 to 14 days needed for germination.

When the seedlings appear, remove the plastic bag and place the trays near a well-lighted window. The best growing temperature is 60 to 70 degrees. Now they will need frequent watering so they will not dry out. When seedlings are two or three inches high, pinch off the less sturdy of the two in each pot, or cut it off with sharp scissors. Don't pull it out or you may damage the roots of the other plant.

A tomato seedling is ready to transplant when it has four or six true leaves (the little "heart" leaves at the bottom of the stem don't count). Lessen the shock of the transition by taking seedlings outside during the day to harden them, and bringing them back in at night, for about a week before transplanting. Begin with a shady spot, and gradually lengthen the time out-of-doors and the time in direct sunlight.

Seedlings can be planted directly in the ground, still in the peat pots or egg cups. Roots will grow through both peat walls and the cardboard, which will disintegrate. If they have grown at all "leggy," plant them deep—more roots will grow all along the buried section of the stem. Or plant at a slant, which keeps more of the stem near the surface where the soil is warmest. They may look funny, but they will straighten up nicely in a few days. Transplant on an overcast day, or shield the plant from too much direct sunlight for the first 48 hours.

You may need to protect the plants on cold nights with plastic bags or hotcaps. If you leave these on during the day, be sure there are ventilation holes or a sunny day can cook the little plant. A large paper grocery sack makes an ideal cover. Cut the closed end of the sack on three sides, leaving a hinged "roof" that can be folded back during the day and closed at night.

If possible, plant seedlings on the south side of your house where they will get maximum sun and shelter from the wind. Planting them next to a cement walkway lets them benefit from reflected and stored heat. For cooler areas, use a cover of aluminum foil or black plastic on the ground just around the plants.

We usually rely only on a heavy mulching around the plants (wasted hay from the barn, that has sheep manure in it) to control weeds, retain moisture, and provide added fertilizer when the plants are watered.

Without pruning, tomato foliage tends to grow too lush, with a scarcity of fruit. Some gardeners pinch off all side branches, leaving a single main stem. Others leave several main branches but remove side growth and pinch off the rapidly-growing suckers that grow out at the juncture of the branches.

Tomatoes love to sprawl, but plants that are staked or framed take less space and the fruit is cleaner. Old nylon stockings, cut into strips or left whole, make ideal soft ties. Or you can make simple lath frames around the plants which will support the branches and fruit without tying. If you make hinged ones, you can store them for next year.

There is also a very effective growing method known as the "Japanese tomato ring." A seven-foot circle of ground is spaded up, with a five-foot circle of sturdy wire screen fencing at least two feet high centered in it. Layers of peat moss and aged manure are piled inside the wire, with a small amount also dug into the foot-wide border outside it. Half a dozen climbing or tall-type tomato plants are planted around the outside of the fence, which will support the plants while their roots feed inside the enriched area. Each year, pile more manure inside the ring before planting again.

I'm kind of a lazy gardener, always intending to try some interesting and elaborate method of tomato culture but each year getting too busy even to prune or tie systematically. Even with this benign neglect, though, the crop is good, so I can enthusiastically recommend tomatoes even if your gardening time is limited. One "system" I always use is to interplant tomatoes with cabbage, and grow some early head lettuce in between. The lettuce is gone by the time the tomatoes and cabbage need more space. The strong smell of the tomato plants in the heat of summer appears to repel or confuse cabbage moths, which flutter in circles above the cabbages but won't land to lay their eggs.

When cool autumn nights slow ripening, cover each plant or the whole area with a loose tent of clear plastic, weighted down with rocks or bricks. The plastic "hothouse" makes it warmer in the daytime and keeps in a lot of the heat during the night. The excess heat and poor ventilation may make the foliage droop, but the tomatoes still ripen.

Tomatoes, both ripe and green, must be picked before the first frost. To ripen them successfully indoors, pick carefully without bruising and leave the stem on. Most people put them on a sunny windowsill, but a dark drawer, a shallow box with a lid, or a brown paper bag works better. Check daily for ripe or spoiled tomatoes. If you have space in your garage, or a cool dry place in your basement, pull up the entire plant when frost is predicted. Hang it, roots and all, upside down, so the tomatoes can ripen on the plant.

But be sure to use many of the green tomatoes **before** they ripen! The very small ones are a nuisance to use ripe anyway, but are ideal for the recipes in this book.

You'll notice that a lot of these recipes, especially in the dessert section, call for "coarsely ground" tomatoes. Finely chopped tomatoes can be substituted, but the result will not be as satisfactory. You may think that grinding them is a lot of bother, and I agree with you, so grind more than you need at one time. They will keep for several days in the coldest part of your refrigerator.

If you have an enormous harvest of green tomatoes, make up a quantity of green tomato mincemeat and can or freeze it, to use through the winter in trying out the 13 recipes using mincemeat. Chutney, jam, marmalade, relishes, chili sauce, etc., are also excellent ways to preserve your bounty. They make unusual and welcome gifts.

Green tomatoes are nutritious and low in calories. One small green tomato has 270 units of provitamin A, 20 milligrams of vitamin C, and only 24 calories. They are ideal for low-sodium diets because they have only 3 milligrams per 100 grams (about 1/2 cup). The salt in most recipes can be omitted or replaced by a low-sodium salt substitute.

BREADS

WINE

GREEN TOMATO QUICK BREAD

VEGETABLE OIL   1/4 cup
SUGAR   3/4 cup
WHOLE EGG   1
LEMON EXTRACT   1/2 teaspoon
GREEN TOMATOES   1 cup ground and undrained
FROZEN CONCENTRATED ORANGE JUICE   2 table-
spoons
FLOUR   2-1/2 cups unsifted (more if tomatoes are juicy)
BAKING POWDER   3 teaspoons
BAKING SODA   1/2 teaspoon
SALT   1/2 teaspoon
POWDERED ORANGE PEEL   1 teaspoon
NUTS or SUNFLOWER SEEDS   1/4 cup chopped
APRICOTS or CANDIED FRUIT   1/4 cup chopped
SUGAR   1/2 teaspoon
POWDERED ORANGE PEEL   1/2 teaspoon

Cream oil, sugar, and egg; beat until fluffy. Add lemon extract, tomatoes, and orange juice; beat. Add dry ingredients, then nuts and fruit. Spoon into 2 greased and floured medium loaf pans. Sprinkle tops with mixture of sugar and orange peel. Bake at 350° for 35 to 40 minutes, or until they test done. Cool in pan for 10 minutes; then turn out onto cake rack.

GREEN TOMATO SPICE BREAD

FLOUR   2-1/2 cups unsifted
BAKING POWDER   2-1/2 teaspoons
BAKING SODA   1 teaspoon
SALT   1/2 teaspoon
BROWN SUGAR   1 cup, firmly packed
CINNAMON   2 tablespoons
NUTMEG   1 teaspoon
GINGER   1/4 teaspoon
MOLASSES   1 tablespoon
HONEY   3 tablespoons
VEGETABLE OIL   1/2 cup
GREEN TOMATOES   1 cup ground and undrained
WHOLE EGGS   2
VANILLA   1 teaspoon
WALNUTS   1/2 cup chopped

Mix dry ingredients. Add rest of ingredients; beat well. Spoon into 2 greased and floured medium loaf pans. Let batter rest 10 minutes. Smooth the top of the loaves; then make an indentation along the center from end to end. Bake at 350° for 45 to 50 minutes, or until they test done.

EASY MINCEMEAT COFFEE RING

FLOUR   2 cups unsifted
BAKING POWDER   2 teaspoons
SALT   3/4 teaspoon
SUGAR   3/4 cup
NUTS   1/2 cup chopped
MILK   5 tablespoons
WHOLE EGG   1
VEGETABLE OIL   1/4 cup
GREEN TOMATO MINCEMEAT   3/4 cup, moist
(page 72-73)
VANILLA   3/4 teaspoon

Mix together flour, baking powder, salt, sugar, and nuts. Add the rest of the ingredients. Stir only until dry ingredients are moistened. Bake in well-buttered 8-inch ring mold at 350° for 30 to 35 minutes. Remove from pan; cool slightly on cake rack, and then frost.

Powdered Sugar Glaze

POWDERED SUGAR   1 cup
MARGARINE   1 tablespoon, softened
VANILLA   1 teaspoon
HOT WATER
NUTS

Mix sugar, margarine, and vanilla together. Beat in enough hot water to make frosting spread easily and dribble over the edge. Sprinkle with nuts.

MUFFINS

FLOUR   1-3/4 cups unsifted

POWDERED MILK   2 tablespoons

BAKING POWDER   3 teaspoons

SALT   1/2 teaspoon

POWDERED LEMON PEEL   1 teaspoon

GREEN TOMATOES   3/4 cup ground and undrained

VEGETABLE OIL   1/4 cup

HONEY   1/3 cup*

WHOLE EGG   1

MILK   1 tablespoon (if needed to moisten)

SUGAR   1 teaspoon

Mix dry ingredients together. Add rest of ingredients and stir just enough to moisten, adding more milk if tomatoes are not juicy. Fill greased muffin pans 3/4 full. Sprinkle tops with sugar. Bake at 400° for 20 to 25 minutes. Makes 12 medium-size muffins.

If honey is measured in the same cup used for oil, it will not stick to cup.

GREEN TOMATO WINE

GREEN TOMATOES   5 pounds, washed and stems
                 removed
RAISINS   2 pounds, ground or chopped
BOILING WATER
WATER   3-3/4 quarts (part BARLEY WATER, optional*)
SUGAR   3-1/2 pounds
ORANGES   2, sliced very thin
WINE YEAST   1/2 packet
YEAST NUTRIENT   1 teaspoon
STRONG TEA   1 cup

Grind or chop tomatoes. Pour boiling water on raisins and let soak for 10 minutes; drain, saving water. Add water to make 3-3/4 quarts. Bring to a boil and dissolve sugar in it. Add tomatoes, raisins, and oranges. When cooled to lukewarm, add wine yeast, softened as directed on package. Stir in yeast nutrient and tea. Cover top of fermenting vessel with cheesecloth or clean tea towel, tied down. Keep in warm place for 3 days, stirring twice a day. Siphon into clean gallon jug, straining out fruit; fit with airlock. Keep in warm place until all signs of fermentation have ceased. Rack off into clean gallon jug, fit with airlock again. Keep in cool place until wine clears. Rack off again and bottle. Recipe quantities may be multiplied for 5 gallons, increasing yeast to one full packet only.

This wine has more body if you add 2 cups barley, soaked and ground. Since grain wines are illegal, I can't recommend barley. However, if you happen to be making barley soup, soak 2 cups barley in water overnight. Then boil 10 minutes. Strain water off and use it, lukewarm, in your wine. Use the barley in your soup.

CAKES

COOKIES

PIES

DESSERTS

EGGLESS GREEN TOMATO CAKE

BROWN SUGAR   1 cup, firmly packed
WATER   1 cup
DATES   1/2 cup chopped
GREEN TOMATOES   1 cup chopped
MARGARINE   1/2 cup
NUTMEG   1/2 teaspoon
CINNAMON   1 teaspoon
VANILLA   1 teaspoon
FLOUR   2 cups unsifted
BAKING POWDER   5 teaspoons
THIN GLAZE FROSTING   (optional)

Mix all but the last 3 ingredients in a large saucepan. Heat and boil about 4 minutes. Cool. Add vanilla. Mix flour and baking powder; add to cooled mixture; beat well. Spoon into greased and floured 9 x 13-inch pan, or into 2 8-inch layer cake pans. Bake at 350° for 30 to 35 minutes, depending on pan size, or until cake tests done. Use a very thin glaze frosting or serve with Hot Mincemeat Sauce.

Hot Mincemeat Sauce

GREEN TOMATO MINCEMEAT   1 cup (page 72-73)
BRANDY   1/2 cup

Heat mincemeat to boiling. Add brandy; stir and reheat, but do not boil. Serve hot over cake.

POOR MAN'S LITTLE FLAT FRUITCAKES

SUGAR   1 cup
RAISINS or CHOPPED DATES   1 cup
GREEN TOMATOES   1 cup coarsely ground and un-
drained
VEGETABLE OIL   1/2 cup
CINNAMON   1 teaspoon
CLOVES   1/2 teaspoon
SALT   1/2 teaspoon
CANDIED FRUIT   1-1/4 cups chopped (or dried fruit)
FLOUR   2 cups unsifted
BAKING SODA   1 teaspoon
NUTS   1/2 cup chopped
BRANDY FLAVORING   2 teaspoons
VANILLA   1 teaspoon
NUTS   slivered (to decorate top)
CORN SYRUP   2 tablespoons (for glaze)
WATER   2 teaspoons

Combine sugar, raisins, tomatoes, oil, and spices in saucepan. Include dried
fruit if using it instead of candied fruit. Bring to boil; remove from heat and
cool. Stir in candied fruit. Mix flour and soda together in bowl. Add fruit
mixture, nuts, and flavorings. Stir until the batter is blended. Spoon into 4
greased and floured 4 x 5 x 2-inch foil pans (or one 7 x 11 x 2-inch baking
pan or muffin pans). Decorate with nuts. Bake the large pan at 300° for 1
hour; bake smaller pans and muffin pans at 275° for about 45 minutes.

To glaze tops, boil corn syrup and water; cool to lukewarm and brush
mixture on the tops of the cakes 15 minutes before they are done. Brush again
after they are done.

Note: These make excellent gifts or can be frozen.

CAKE WITH GREEN
TOMATO FROSTING-FILLING

FLOUR   1-3/4 cups unsifted
BAKING POWDER   2 teaspoons
SUGAR   1 cup
VEGETABLE OIL   1/4 cup
WHOLE EGG   1
MILK   3/4 cup
GREEN TOMATOES   3 tablespoons ground
BUTTERSCOTCH FLAVORING   1 teaspoon

Mix dry ingredients; add rest of ingredients; beat. Bake in 2 greased and floured 8-inch layer cake pans at 350° for 20 to 25 minutes, or until done.

Frosting-Filling

GREEN TOMATOES   1 cup chopped
SUGAR   1/2 cup
UNFLAVORED GELATIN   1 tablespoon (1 envelope)
COLD WATER   2 tablespoons
FROZEN CONCENTRATED ORANGE JUICE   1/2 cup
NUTS   1/2 cup chopped

Mix tomatoes and sugar. Heat slowly, simmering until tender. Mix briefly in blender. Soften gelatin in cold water; add to hot tomato mixture, and stir until well dissolved. Stir in orange juice. Cool until mixture begins to set. Spread thinly over each layer of cooled cake. Place layers in refrigerator for 10 minutes. Take them out; spread thick layer of gelatin mixture over bottom layer, top with other layer; spread thin layer over whole cake. Sprinkle with chopped nuts. Chill until ready to serve.

MINCEMEAT CUPCAKES

MARGARINE   1/3 cup
SUGAR   1/3 cup
WHOLE EGGS   2
VANILLA   1 teaspoon
GREEN TOMATO MINCEMEAT   1 cup (page 72-73)
SOUR MILK*   1/3 cup
FLOUR   2 cups unsifted
BAKING POWDER   1-1/2 teaspoons
BAKING SODA   1/2 teaspoon

Cream margarine and sugar; add eggs; beat until fluffy. Stir in vanilla, mincemeat, and sour milk. Add other ingredients and mix well. Fill muffin pans, lined with fluted paper cups, 2/3 full. Bake at 375° for 20 to 25 minutes. Makes 24 medium, or 12 to 16 large cupcakes.

*To make sour milk, add 1/2 teaspoon lemon juice or vinegar to 1/3 cup milk. Let stand 5 minutes before using.

LAYERED DESSERT

Filling

GREEN TOMATOES   2 cups chopped
HONEY   2 tablespoons
CORNSTARCH   3 tablespoons
COLD WATER   3 tablespoons
BROWN SUGAR   1 cup, firmly packed
VANILLA   1 teaspoon

Simmer tomatoes and honey for 5 minutes. Add cornstarch dissolved in cold water; add brown sugar. Simmer until thick. Add vanilla. Cool.

Layers

QUICK COOKING OATS   1-1/2 cups
BROWN SUGAR   1 cup, firmly packed
VEGETABLE OIL   7/8 cup
FLOUR   1-1/2 cups unsifted
BAKING POWDER   1 teaspoon
BAKING SODA   1/2 teaspoon
SUNFLOWER SEEDS   3/4 cup chopped (in blender)
WHIPPED CREAM

Mix oats, sugar, and oil. Mix the dry ingredients together; add to oil and sugar mixture and mix well. Stir in sunflower seeds. Spread half of this mixture in greased 9 x 9-inch pan. Pour tomato filling over it, and top with other half of oatmeal layer. Bake at 350° for 45 minutes. Serve warm with whipped cream. Serves 9 to 12.

TURNOVERS

Crust

FLOUR   2 cups unsifted
MARGARINE   1/2 cup
BUTTERMILK   1/4 cup

Cut margarine into flour. Stir in buttermilk to form dough. Roll out on pastry cloth. Cut into triangles or 3-inch rounds.

Filling

GREEN TOMATOES   1 cup chopped
HONEY   1/3 cup
CORNSTARCH   4 teaspoons
COLD WATER   1 tablespoon
SOY GRITS   1 tablespoon
MARGARINE   1 tablespoon
SHREDDED COCONUT   1/3 cup
BUTTERMILK
SUGAR or POWDERED ORANGE PEEL

Mix tomatoes and honey together. Bring to boil and simmer 5 minutes. Dissolve cornstarch in cold water. Add to tomato mixture along with soy grits. Simmer and stir until thickened and clear. Stir in margarine and coconut. Cool.

Put scant tablespoon of filling on each round or triangle. Moisten edges with buttermilk; seal. Prick tops with fork. Brush with buttermilk, and sprinkle lightly with sugar or orange peel. Place on baking sheet and bake at 400° for 18 to 20 minutes, until light golden brown.

MINCEMEAT-STRETCHER PUDDING

APPLES   3-1/2 cups peeled and sliced
FROZEN CONCENTRATED APPLE JUICE   1/2 cup, plus
                                  3 tablespoons thawed
SUGAR   1/2 cup (if using Martha's Vineyard Mincemeat)
CORNSTARCH   1 tablespoon
MARGARINE or VEGETABLE OIL   1 tablespoon
GREEN TOMATO MINCEMEAT   2 cups (page 72-73)
WHIPPED CREAM

Cook apples slowly in 1/2 cup apple juice, stirring often. (If using the
Martha's Vineyard Mincemeat recipe, add sugar.) When tender, add cornstarch dissolved in 3 tablespoons of apple juice. Cook, stirring until thickened; then cook slowly until translucent. Add margarine and mincemeat; stir
until margarine is melted. Spoon into dessert dishes or sherbert glasses.
Serve slightly warm with sweetened whipped cream. Serves 8.

MINCEMEAT PINWHEELS

MARGARINE   1/4 cup
BROWN SUGAR   1 cup, firmly packed
WHOLE EGG   1
VANILLA   1 teaspoon
FLOUR   2 cups unsifted
BAKING SODA   1/2 teaspoon
BAKING POWDER   1/2 teaspoon
CINNAMON   1/4 teaspoon

Cream margarine and sugar. Add egg and beat well. Stir in vanilla, and then dry ingredients, mixed together. Press dough into a ball; wrap in plastic wrap. Chill for an hour or more.

Filling

GREEN TOMATO MINCEMEAT   1-1/2 cups, thick (page 72-73)
NUTS or SUNFLOWER SEEDS   1/2 cup chopped
MARGARINE   1 tablespoon, melted
WHEAT GERM   2 tablespoons (more if mincemeat not thick)
FINELY SHREDDED COCONUT   1/4 cup
SUGAR   1/4 cup (if using Martha's Vineyard Mincemeat)
POWDERED SUGAR

Mix together mincemeat, nuts, margarine, wheat germ, and coconut. (If using Martha's Vineyard Mincemeat recipe, add sugar.) Roll out chilled dough on pastry cloth to a rectangle about 1/4-inch thick. Spread mincemeat filling on dough. Roll up, starting at the long edge. Wrap roll in plastic; chill in refrigerator for several hours or until next day. Cut in 1/2-inch slices. Bake on lightly greased baking sheet at 350° for 20 to 25 minutes. Dust with powdered sugar.

MINCEMEAT IN SNOW

HOT WATER   1/2 cup
UNFLAVORED GELATIN   1 tablespoon (1 envelope)
VEGETABLE OIL   1/4 cup
HONEY   1/4 cup
VANILLA   1 teaspoon
ICE CUBES   4 to 6, crushed
GREEN TOMATO MINCEMEAT   1 cup (page 72-73)

Put hot water in blender container; add gelatin and blend. With blender running, add oil, honey, and vanilla. With blender running, add crushed ice cubes, one at a time, until mixture starts to thicken. Pour into bowl at once; fold in mincemeat. Spoon into sherbet or parfait glasses. This should be solid enough to serve within 5 minutes. Refrigerate if not serving soon. Serves 4.

MINCEMEAT FLUFF

EGG WHITES   4
SALT   1/8 teaspoon
SUGAR   1/3 cup
GREEN TOMATO MINCEMEAT   3/4 cup, thick
(page 72-73)
LEMON JUICE   1 teaspoon
WHIPPED CREAM

Beat egg whites until foamy; add salt. Beat until barely stiff. Add sugar gradually, beating until stiff. Fold in mincemeat and lemon juice. Spoon into 1-quart buttered casserole or custard cups. Bake at 300° for 25 to 35 minutes, or until firm. Serve with whipped cream.

MINCEMEAT COBBLERS

Cobbler Crust

FLOUR   2 cups unsifted
SUGAR   3 tablespoons
SALT   1/2 teaspoon
BAKING POWDER   4 teaspoons
MARGARINE   6 tablespoons
MILK   3/4 cup
VANILLA   few drops

Filling

GREEN TOMATO MINCEMEAT (page 72-73)
FRUIT JUICE or SUGAR-WATER

Mix dry ingredients together. Cut in margarine until like coarse meal. Stir in milk and vanilla to make a soft dough. Roll out on pastry cloth with cloth-covered rolling pin to thickness of pie crust. Cut into large circles. (Freeze extra cobbler crusts, using 2 circles of waxed paper between each crust. These defrost quickly because they are so thin.) Butter small oven-proof bowls. Spoon 1/3 to 1/2 cup mincemeat into each bowl. Heat until bubbly in a 400° oven. Place cobbler crust on top of each; brush top of crust with a bit of fruit juice or sugar-water. Return to oven for 8 to 10 minutes, until lightly browned. Serve slightly warm.

CHESS PIES

Make filling before crust so that it can cool.

Filling

>   GREEN TOMATOES   2/3 cup coarsely ground and un-
>   drained
>   DATE CRYSTALS*   1 cup or
>   DATES   1 cup chopped and CORNSTARCH 1 teaspoon
>   BROWN SUGAR   1 cup, firmly packed
>   MARGARINE   1/2 cup
>   EGG YOLKS   3
>   VANILLA   1 teaspoon
>   NUTS   3/4 cup chopped

Mix tomatoes, dates, and cornstarch. Cook together for 10 minutes. Cool. Cream brown sugar, margarine, and egg yolks. Add vanilla and date mixture; beat well. Stir in chopped nuts.

*Order date crystals from Shield's Date Gardens, 88-225 Highway 111, Indio, CA 92201.

Pastry

>   FLOUR   1-1/2 cups unsifted
>   BAKING POWDER   1/2 teaspoon
>   SUGAR   1 tablespoon
>   MARGARINE   1/2 cup, melted
>   COLD WATER   1/4 cup
>   MAKE-AHEAD WHIPPED CREAM or MERINGUE

Combine flour, baking powder, and sugar. Add melted margarine to water and stir into dry ingredients. Mix to a smooth dough. Roll out on well-floured pastry cloth to thin layer. Cut into 3-inch circles. Reroll scraps and cut more circles for a total of about 18. This is a soft pastry, easy to fit into muffin pans.

*Fit pastry circles into cups of muffin pans. Put 1 heaping tablespoon of filling
in each pastry-lined cup. Bake at 350° for 25 to 28 minutes until centers and
crust are golden brown. Cover with meringue, if desired, and put back in
oven to brown. Or serve with Make-Ahead Whipped Cream.*

Make-Ahead Whipped Cream

UNFLAVORED GELATIN   1 teaspoon
COLD WATER   3 tablespoons
WHIPPING CREAM   1 cup, chilled
SUGAR   2 tablespoons
VANILLA   1/2 teaspoon

*Soak gelatin in cold water; heat carefully to dissolve. Add to whipping cream
in bowl. Let chill 1 hour in refrigerator to partially set. Add sugar and vanilla;
whip stiff.*

*This will keep its shape for several hours without refrigeration. In the
refrigerator it will keep for several days. Stir before using, or scoop out with
ice cream scoop without stirring.*

CANDIED GREEN TOMATO PEEL

LARGE GREEN TOMATOES   4, unpeeled
SUGAR   1/2 cup
WATER   1 tablespoon
CORN SYRUP   1 tablespoon
RED VEGETABLE COLORING   4 drops (optional)

*Slice tomatoes; remove seeds and soft pulp. Chop firm flesh into small
cubes. Drain well on paper towels. Boil up sugar, water, and corn syrup. Add
red coloring, if desired. Add tomato cubes and cook until almost transparent
(240° —250° on candy thermometer). Remove tomato cubes with slotted
spoon and spread on foil (not wax paper) to dry. Drying can be speeded by
putting them in a warm oven for several hours.*

GREEN TOMATO PIE
(or cobblers)

GREEN TOMATOES   4 cups halved and sliced 1/4-inch
thick
LEMON JUICE   1 tablespoon
SUGAR   1 cup
SALT   1/2 teaspoon
CINNAMON   1 teaspoon
POWDERED ORANGE PEEL   2 teaspoons
CORNSTARCH   2 tablespoons
BUTTER or MARGARINE   1 tablespoon
PASTRY   for 9-inch double crust pie, or cobbler crusts

Sprinkle tomatoes with lemon juice. Combine dry ingredients; mix lightly with tomatoes.

For pie: Spoon tomato mixture into 9-inch pastry lined pie pan. Dot with butter. Cover with top crust; seal edges. Cut slits in top. Bake at 400° for 10 minutes. Then lay aluminum foil loosely across top of pie. Bake at 350° for 40 to 50 minutes longer, until filling is tender and thickened.

For cobbler: Spoon tomato mixture into buttered oven-proof bowls. Bake at 350° for 40 minutes, stirring once after 20 minutes. Then turn heat up to 425°. Top tomato mixture with cobbler crusts (see recipe for Mincemeat Cobblers) moistened on top with a bit of the hot juice of the tomato mixture. Bake at 425° 8 to 10 minutes until crusts are pale golden brown.

GREEN TOMATO PIE
(unbaked filling)

HONEY   3/4 cup
APPLE JUICE   2/3 cup
CORNSTARCH   3 tablespoons
WATER   4 tablespoons
VERY SMALL GREEN TOMATOES   1 quart halved
MARGARINE or BUTTER   1 tablespoon
BAKED PIE SHELL   9-inch
WHIPPED CREAM

Boil up honey and apple juice. Add cornstarch dissolved in water; cook until thickened. Add tomatoes and cook until tender, being careful not to mash them. Add margarine and stir in when it is melted. Cool this mixture. Then pour into baked 9-inch pie shell. Refrigerate until time to serve. Top with whipped cream.

MAGIC TOMATO PIE (crustless)

SUGAR   3/4 cup

WHOLE EGG   1

FLOUR   6 tablespoons

SALT   1/4 teaspoon

BAKING POWDER   1-1/2 teaspoons

POWDERED ORANGE PEEL   1 teaspoon

GREEN TOMATOES   1 cup coarsely ground, drained, and
squeezed dry

VANILLA   1 teaspoon

SUNFLOWER SEEDS or PECANS   1/2 cup chopped

WHIPPED CREAM

Mix all ingredients together. Pour into heavily buttered 9-inch pie pan. Bake at 350° for 45 minutes; cover loosely with aluminum foil for the last 10 or 15 minutes if it starts getting too brown. Serve with whipped cream.

TOMATO LATTICE PIE

PASTRY for 9-inch lattice-top pie

Line 9-inch pie pan with pie dough, and assemble lattice top on wax paper. Refrigerate both while preparing pie filling.

Filling

TAPIOCA   1 tablespoon
SUGAR   1/2 cup
ALLSPICE   1/4 teaspoon
SALT   1/4 teaspoon
HONEY   2 tablespoons
LEMON JUICE   1 tablespoon
MARGARINE   3 tablespoons, melted
GREEN TOMATOES   4 cups sliced (discard thin slice from top and bottom of tomato)
CHEESE

Sprinkle pie crust with tapioca. Mix dry ingredients together. Mix honey, lemon juice, and margarine together. Layer pie crust alternately with half of the tomatoes, half of the dry ingredient mixture, and half of the honey mixture. Repeat with remaining half of mixtures. Put on lattice top and crimp edges. Bake at 375° for 45 minutes, covering loosely with foil for the last 15 minutes if the top is getting too brown. Serve with cheese.

POVERTY COOKIES

FLOUR   2 cups
SUGAR   1-1/3 cups
SALT   1/2 teaspoon
BAKING SODA   1 teaspoon
BAKING POWDER   1 teaspoon
CINNAMON   1 teaspoon
POWDERED LEMON PEEL   1/2 teaspoon
SOY GRITS   3 tablespoons
QUICK COOKING OATS   2 cups
GREEN TOMATOES   1 cup ground and drained (reserve juice)
VEGETABLE OIL   2/3 cup
MILK plus GREEN TOMATO JUICE   to make 2/3 cup
VANILLA   1 teaspoon

Mix dry ingredients. Add rest of ingredients. (May need more flour if tomatoes are very juicy.) Mix. Drop from teaspoon on greased cookie sheet. Bake at 350° for 15 minutes. Makes 5 dozen.

SQUARE GREEN TOMATO PIE

Crust and Topping

FLOUR   2 cups unsifted
SUGAR   3 tablespoons
MARGARINE   1 cup
BROWN SUGAR   3 tablespoons (for topping)

Mix flour and sugar together; add sliced margarine. Mix with pastry blender until it resembles fine meal, or mix at lowest speed of electric mixer. Divide into thirds. Pat and spread 1/3 of it on bottom only of a 9 x 9 x 2-inch pan. Pat 1/3 of it into little tart shell pans and bake for later use. Mix brown sugar with rest of mixture; crumble this in the bottom of a pie pan. Bake at 350° for 8 to 12 minutes until a pale golden color. Do not overbake. (Tart shells will have to be taken out first.) Stir the crumbly topping occasionally as it bakes. Cool. These can be made a day ahead.

Filling

GREEN TOMATOES   2 cups chopped
HONEY   2/3 cup
CORNSTARCH   2-1/2 tablespoons
COLD WATER   3 tablespoons
SOY GRITS   1 tablespoon
MARGARINE   1 tablespoon
VANILLA   1 teaspoon
POWDERED ORANGE PEEL   1 teaspoon
WHIPPED CREAM

Mix tomatoes and honey; bring to boil and simmer for 5 minutes. Stir in cornstarch dissolved in cold water. Simmer until thick and tomato is tender. Stir in soy grits, margarine, vanilla, and orange peel. Cool. This can be made the day before.

To assemble: *Spoon filling over baked 9 x 9-inch crust; top with sweetened whipped cream and sprinkle the crumbled crust mixture over this. Refrigerate until ready to use. Cut in squares.*   33

MINCEMEAT COOKIES

MARGARINE   1 cup
BROWN SUGAR   2 cups, firmly packed
WHOLE EGGS   2
FLOUR   3-1/2 cups unsifted
SALT   1 teaspoon
BAKING POWDER   1-1/2 teaspoons
BAKING SODA   1/2 teaspoon
MILK   1/4 cup
GREEN TOMATO MINCEMEAT   2 cups (page 72-73)
NUTS or SUNFLOWER SEEDS   1 cup chopped

Cream margarine and sugar; beat until fluffy. Add eggs and beat. Mix dry ingredients together and add alternately with milk. Stir in mincemeat and nuts. Drop from teaspoon on greased cookie sheet. Bake at 400° for 10 to 12 minutes, or until lightly browned. Makes 8 or 9 dozen. These freeze well.

MINCEMEAT-FILLED COOKIES

MARGARINE   1 cup
BROWN SUGAR   1 cup, firmly packed
SOUR MILK*   1/2 cup
QUICK COOKING OATS   4 cups
SALT   1/2 teaspoon
BAKING SODA   1 teaspoon
FLOUR   scant 2 cups unsifted
NUTMEG   1/2 teaspoon
GREEN TOMATO MINCEMEAT   1 pint, for filling (page 72-73)

Cream margarine and sugar. Add sour milk, then oats. Mix rest of ingredients together, except mincemeat. Add to creamed mixture and mix well. Chill 15 minutes. Roll out cookie dough on pastry cloth to thin layer. Cut in 3-inch circles. Place 1 teaspoon of filling on half the cookie circles; top with another cookie circle, and press edges together with a fork. Bake on greased cookie sheet at 350° for 12 to 15 minutes. Cool on rack.

*To make sour milk, add 1 teaspoon lemon juice or vinegar to 1/2 cup milk. Let stand 5 minutes before using.

COFFEE COOKIES

FLOUR   2 cups unsifted
INSTANT COFFEE   1 to 2 tablespoons powdered (depending on taste)
SALT   1/2 teaspoon
BAKING SODA   1/2 teaspoon
SUNFLOWER SEEDS   1/2 cup chopped
VEGETABLE OIL   6 tablespoons
VANILLA   1 teaspoon
WHOLE EGGS   2
SUGAR   1-1/4 cups
GREEN TOMATOES   3/4 cup ground and drained

Mix flour, coffee, salt, soda, and sunflower seeds. Add remaining ingredients; mix. Drop from teaspoon on greased cookie sheet. Bake at 350° for 15 minutes. Makes 5 dozen.

BAR COOKIES

Bottom Layer

FLOUR   1 cup unsifted
SUGAR   2 tablespoons
MARGARINE   1/2 cup

Mix flour and sugar; blend in margarine with pastry blender, or at low speed with electric mixer. Pat this into 9 x 13 x 2-inch pan. Bake at 350° for 8 minutes.

Top Layer

WHOLE EGGS   2
BROWN SUGAR   3/4 cup, firmly packed
GREEN TOMATOES   1/2 cup coarsely ground and slightly drained

SHREDDED COCONUT   1 cup
POWDERED ORANGE PEEL   2 teaspoons

Mix all ingredients together and pour onto baked bottom layer. Bake at 350° for 20 to 25 minutes, until golden brown.

FILLED COOKIES

MARGARINE   1 cup
SUGAR   2 cups
WHOLE EGGS   3
FLOUR   4 cups unsifted
BAKING SODA   1/2 teaspoon
SALT   1/4 teaspoon
VANILLA   1 teaspoon

Cream margarine and sugar; add eggs and mix well. Add rest of ingredients and mix. Roll very thin on floured pastry cloth. Cut in small triangles; transfer to ungreased cookie sheet.

Filling

GREEN TOMATOES   3/4 cup ground and drained
SUGAR   3/4 cup
CORNSTARCH   4 tablespoons
COLD WATER   1/4 cup
POWDERED ORANGE PEEL   1 teaspoon
MARGARINE   1 tablespoon

Put tomatoes and sugar in saucepan. Dissolve cornstarch in water, and add. Add orange peel. Cook until thickened. Stir in margarine. Cool. Then put 1/2 teaspoon of filling on each cookie triangle. Turn up points to center of cookie; press together. Bake at 350° for 10 minutes until delicately browned. Do not overbake.

MINCEMEAT CHOCOLATE CAKE

SUGAR   3/4 cup (increase to 1 cup if using Martha's Vineyard
              Mincemeat)

VEGETABLE OIL   3 tablespoons

WHOLE EGGS   2

GREEN TOMATO MINCEMEAT   1-1/2 cups (page 72-73)

FLOUR   1-7/8 cups unsifted

COCOA   3 tablespoons

BAKING SODA   1 teaspoon

BAKING POWDER   1/2 teaspoon

Cream sugar, oil, and eggs together. Beat until fluffy. Stir in mincemeat, then
dry ingredients. (If mincemeat is unusually thick, you may need to add 1
tablespoon apple juice, to make better consistency.) Bake in a greased and
floured 9 x 9-inch pan at 350° for about 35 to 40 minutes, or until cake tests
done. Cool. Good with sweetened yogurt, or with Fluffy Caramel Frosting.

Fluffy Caramel Frosting

FLOUR   2-1/2 tablespoons

WHOLE MILK or SKIM MILK   1/2 cup

BROWN SUGAR   1/2 cup, firmly packed

MARGARINE   1/2 cup, soft

VANILLA   1-1/2 teaspoons or

MAPLEINE   1 teaspoon

Shake flour and milk in small covered jar to blend. Pour into small saucepan
and cook over low heat until it is thick, stirring to keep it from sticking. Set it
aside to cool. Cream sugar and margarine until light and fluffy. Add
flavoring. Add cooled flour mixture and beat until frosting is like whipped
cream. Spread on cake, swirling the top.

Chocolate Fluffy Frosting

Using same recipe, but add 3 tablespoons chocolate chips to the hot cooked
flour mixture; stir to melt. Reduce margarine to 1/3 cup.

39

CHERRY TOMATO CLAFOUTI*

Batter

WHOLE EGGS   3
FLOUR   1-1/2 cups unsifted
BEER   3/4 cup
SALT   pinch
SUGAR   1 teaspoon
BUTTER   1/2 cup (or VEGETABLE OIL)

Mix together eggs, flour, beer, salt, and sugar. Heat the butter until it is barely hazelnut colored (do not burn it) and pour it hot on the batter; mix. Let batter rise in warm place for 3 or 4 hours.

FRESH RIPE CHERRIES   1 cup pitted and cut in half
GREEN CHERRY TOMATOES   1 cup cut in fourths (or in half if very small)

Syrup

SUGAR   1-1/2 cups
WATER   1/2 cup
FROZEN CONCENTRATED ORANGE JUICE   1 table-spoon

LEMON   1, juice and grated rind

Prepare cherries and drain. Cut tomatoes in quarters; shake out the seeds and juice. Poach tomatoes in sugar-water-orange-lemon syrup until barely tender; drain well. Reserve syrup to pour over clafouti when serving.

*As served by Chef Francois Kissel at the Brasserie Pittsbourg, Seattle.

SUGAR   6 tablespoons
BUTTER

Shake drained tomatoes and cherries in 1/4 cup of sugar to coat. Arrange them in buttered 8-inch cake pan, or pie pan, which has been sprinkled with 1 tablespoon sugar. (Or fruits can be folded into batter.) Stir batter; pour over fruit. Bake at 400° for 1/2 hour. Sprinkle with 1 tablespoon sugar then continue to bake for about 10 minutes, until cake tests done. Good served warm or cold with syrup poured over it.

Author's Note: For more leavening, add to beer mixture 1 teaspoon of granulated yeast softened in 1 tablespoon lukewarm water.

"Custard Clafouti"
(Author's alternate batter)

WHOLE EGGS   3
SUGAR   1/2 cup
FLOUR   1/2 cup unsifted
SALT   pinch
VEGETABLE OIL   3 tablespoons
MILK   1-1/4 cups
VANILLA   1/2 teaspoon

Beat together and pour over fruits in 9-inch pan.

MINCEMEAT ICE CREAM

UNFLAVORED GELATIN   1 teaspoon
COLD WATER   2 tablespoons
EVAPORATED CANNED MILK   1/4 cup
HONEY   1/3 cup (use 1/2 cup if using Martha's Vineyard
        Mincemeat)
CREAM   1-1/4 cups
VANILLA   3/4 teaspoon
LIQUID LECITHIN   1 tablespoon (optional)
GREEN TOMATO MINCEMEAT   3/4 cup (page 72-73)

Soften gelatin in cold water. Add milk; heat to dissolve gelatin. Add honey;
heat to liquify. Stir well. Combine with cream and vanilla. Chill mixture for 2
to 3 hours. Pour into Ice Cream Machine; add lecithin, and freeze according
to machine directions for 1/2 hour. Remove Ice Cream Machine from freezer,
add mincemeat, and return to freezer until ice cream is done and machine
shuts off. Makes approximately 1 quart. Good plain or with:

Mincemeat Sauce

GREEN TOMATO MINCEMEAT   1 cup (page 72-73)
HONEY   2 tablespoons
BRANDY   1/4 cup

Combine all ingredients and serve with the ice cream.

Note: Recipe for use in Salton Ice Cream Machine.

MINCEMEAT BAR COOKIES

MARGARINE   1/2 cup, soft
HONEY   1 cup
VANILLA   1/2 teaspoon
POWDERED ORANGE PEEL   1/2 teaspoon grated
WHOLE EGGS   3
BAKING POWDER   1 teaspoon
FLOUR   1-1/4 cups unsifted
WHEAT GERM   2 tablespoons
GREEN TOMATO MINCEMEAT   1 cup (page 72-73)
WALNUTS   1 cup broken

Beat margarine and honey together until fluffy. Add vanilla and orange peel.
Beat in eggs, one at a time. Add dry ingredients and stir just until mixed. Stir
in mincemeat and nuts. Bake in greased and floured 9 x 13-inch pan at 350°
for about 25 to 30 minutes. While hot, frost top with very thin dribble of
Powdered Sugar Glaze; cut into bars.

ICE CREAM SUNDAE TOPPING*

HONEYDEW MELON   *2 cups, diced or in small balls, or both*

GREEN TOMATOES   *2 cups sliced*

SUGAR   *2 cups*

LEMON   *1, juice and grated rind or* **LEMON**   *1, very thinly sliced*

CINNAMON STICK   *1 (optional)*

Combine melon, tomatoes, sugar, and lemon juice and rind; cook together until thick, flavoring with cinnamon when almost done, if desired. Pour into hot washed half-pint jars; seal at once.

Good as ice cream topping, or cake filling.

*As served by Chef Francois Kissel at the Brasserie Pittsbourg, Seattle.

CASSEROLES

BRUNCH EGGS

BACON   4 slices
MEDIUM ONION   1, sliced thin
MEDIUM-SIZE GREEN TOMATOES   2, sliced very thin
WHOLE EGGS   6
DAIRY SOUR CREAM   1/2 cup
DRY MUSTARD   1 teaspoon
SALT   1/2 teaspoon
FRESH GROUND PEPPER   1/4 teaspoon

Fry bacon crisp; remove from pan. Pour off all but 2 tablespoons fat. Saute onion and tomatoes in bacon drippings until limp and tender. Add remaining ingredients. Scramble over low heat. Top with crumbled bacon. Serves 4.

BRUNCH EGGS PLUS

RIPE TOMATOES   3, quartered
MEDIUM-SIZE GREEN TOMATOES   2, halved and
sliced thin

SMALL ONION   1/2, chopped
MARGARINE   2 tablespoons
FLOUR   2 tablespoons
CREAMED CORN   1 15-ounce can
WHOLE EGGS   2, beaten
SHARP CHEDDAR CHEESE   3/4 cup shredded
SALT and PEPPER   to taste
COOKED NOODLES   hot and buttered

*Saute vegetables in margarine until tender. Stir in flour and corn. Cook until
thickened. Add eggs and cheese and season to taste. Stir until eggs are set.
Serve on buttered noodles. Serves 4.*

POTATO CASSEROLE PIE

MEDIUM-SIZE GREEN TOMATOES   4, halved and
sliced thin

SUGAR   1/2 teaspoon

SEASONED SALT   1/2 teaspoon

SMALL ONION   1, chopped

CUCUMBER   1, thinly sliced

BUTTER or MARGARINE   1 tablespoon

HOT MASHED POTATOES   1-1/2 cups, well-seasoned

SHARP CHEDDAR CHEESE   1/2 cup grated

Arrange tomato slices overlapping in bottom of oiled casserole or large pie pan. Sprinkle with sugar and half of the salt. Layer onion and cucumber over tomatoes, and sprinkle with rest of salt. Dot with butter. Bake at 350° for 15 minutes. While this bakes, prepare potatoes. Top the casserole with potatoes and sprinkle with cheese. Bake at 375° until browned. Serves 4.

JIFFY GREEN TOMATO PIZZAS

FLOUR   2 tablespoons
TOMATO SAUCE   1 8-ounce can
TOMATO PASTE   1 6-ounce can
LARGE GREEN TOMATOES   2, chopped
GARLIC CLOVES   2, crushed
OREGANO   1 teaspoon crushed
BASIL   1/4 teaspoon crushed
ONION   1/4 cup chopped
HOT ITALIAN SAUSAGE   1 (optional)
ENGLISH MUFFINS   6, split
MARGARINE
MOZZARELLA CHEESE   12 slices
RIPE OLIVES   sliced (for garnish)
MUSHROOMS   sliced (for garnish)

Stir flour into tomato sauce until smooth. Add tomato paste, tomatoes, garlic, oregano, basil, and onion. Simmer until thick and vegetables are tender. While sauce cooks, slice sausage very thin, and saute in skillet without oil until lightly browned on both sides. Remove to absorbent paper to drain. Spread muffin halves with margarine. Spoon the sauce onto muffins, covering tops. Place slice of cheese on each; then decorate with sausage, olives, and mushrooms. Put under broiler until cheese is melted and bubbly. Serves 6.

TORTILLA CASSEROLE

LEAN GROUND BEEF   1 pound
VEGETABLE OIL   1 tablespoon
ONION   1/2 cup chopped
FLOUR   2 tablespoons
TOMATO SAUCE   1 15-ounce can
LARGE GREEN TOMATOES   3, chopped
RIPE TOMATO   1, chopped
GARLIC CLOVES   2, crushed
GREEN CHILI SALSA   1/2 cup (optional)
CHILI POWDER   1 tablespoon
CUMIN   1 teaspoon
SALT   1/2 teaspoon or more
FRESH GROUND PEPPER   1/4 teaspoon
SUGAR   1/2 teaspoon
WATER   1 cup
TORTILLAS   12, homemade or purchased
SHARP CHEDDAR CHEESE   3/4 pound, shredded (about
3 cups)
ONION   3/4 cup chopped

Brown beef in oil; drain off excess fat. Add 1/2 cup onion and saute until limp.
Stir in flour; blend in tomato sauce. Add both ripe and green tomatoes,
garlic, chili salsa (if desired), and seasonings. Stir in the water. Simmer sauce
slowly until thickened.

Fry tortillas lightly in oil. Spoon 1 tablespoon cheese, 1 tablespoon raw
onion, and 2 tablespoons of meat sauce onto each tortilla. Fold over and
arrange filled tortillas in large shallow baking pan, side by side. Spoon
remaining meat sauce over them; sprinkle with remaining cheese. Bake at
350° for 25 to 30 minutes, until cheese is bubbly. Serves 6.

Easy Home-Made Tortillas

FLOUR   1-1/4 cups
YELLOW CORNMEAL   3/4 cup
SALT   3/4 teaspoon
BOILING WATER   1 cup
VEGETABLE OIL   2 tablespoons

Mix dry ingredients together. Boil up oil in water; add dry ingredients and mix well. Divide dough into 12 balls. Roll each into a thin disk on floured pastry cloth or between wax paper.

In lightly oiled skillet, brown tortillas on each side. These can be made ahead and reheated in oiled skillet to soften them when ready to use. Makes 12 tortillas. Recipe can be doubled.

BACON-AND-TOMATO LUNCH

MEDIUM-SIZE GREEN TOMATOES   8 (2 per person)
BACON   16 thin slices
SALT and PEPPER
PREPARED MUSTARD   1/4 teaspoon per tomato
BROWN SUGAR
FLUFFY COOKED RICE

Slice off top of each tomato; shake out the seeds. Lightly saute bacon until partially cooked. Criss-cross 2 slices of bacon and put a tomato in the center. Sprinkle with salt and pepper; spread with mustard. Close bacon ends over the top and fasten with toothpick. Repeat for each tomato. Place on rack in broiler pan. Sprinkle lightly with brown sugar. Bake at 450° for 20 minutes. Bacon should be crisp and tomatoes hot. Serve on bed of hot rice. Serves 4.

SPANISH RICE

LARGE GREEN TOMATOES   3, cut in chunks
WATER   2-1/4 cups
ONION   1, cut in several pieces
BROWN RICE   1 cup
CHICKEN BOUILLON CUBES   2
GARLIC CLOVES   2, minced or pressed through garlic
press
SEASONED SALT   1/2 teaspoon or more
HOT PEPPER SAUCE   few drops
MOLASSES   1 teaspoon
RIPE TOMATOES   2 cups chopped
PARSLEY   1 sprig, chopped
TOMATO SAUCE   1/2 cup
BACON   4 slices
GREEN PEPPER   1/2, chopped
PARMESAN CHEESE   grated

Place green tomatoes and water in blender container, and blend. Add onion, blending only enough to chop it. Heat this mixture to boiling; add rice, bouillon, garlic, salt, pepper sauce, and molasses. Simmer, covered, until rice has absorbed most of the water. Add ripe tomatoes, parsley, and tomato sauce. Fry bacon crisp; drain it on paper towel. Saute green pepper in bacon fat; add it to rice. Crumble bacon; add half of it to rice. Mix in. Turn rice into oiled 1-quart casserole, top with rest of the crumbled bacon. Bake at 350° for 20 minutes, or until rice is tender. Serve with grated Parmesan cheese. Serves 4.

SAUCY STEW MEAT

GREEN TOMATOES   4
WATER   1/3 cup
SEASONED SALT   1/2 teaspoon
SOY SAUCE   1 tablespoon
GARLIC CLOVES   2
BEEF STEW MEAT   1 pound
RIPE TOMATOES   3, quartered
ONION   1 cup chopped
PARSLEY   1/2 cup chopped
MUSHROOMS   1/2 cup sliced (or 1 small can, drained)
BUTTERED NOODLES

In blender, combine green tomatoes, water, salt, soy sauce, and garlic; blend well. Brown meat in Dutch oven. Add tomato mixture, ripe tomatoes, and onion. Cover and simmer 1-1/2 hours. Add parsley and mushrooms. Simmer uncovered 15 minutes to reduce and thicken the sauce. Serve over buttered noodles. Serves 4.

LITTLE SAUSAGE CASSEROLES

> *BROWN-N-SERVE or PORK LINK SAUSAGES*   *16*
> *POTATOES*   *2 cups cooked and sliced*
> *SMALL GREEN TOMATOES*   *4, thinly sliced*
> *ONION*   *4 thin slices*
> *SEASONED SALT*
> *FRESHLY GROUND PEPPER*
> *EVAPORATED MILK*   *1 cup*
> *DROP-BISCUIT DOUGH with grated SHARP*
>   *CHEDDAR CHEESE (2-cup recipe)*

Fry and drain sausages. Place 4, cut in half, in bottom of each individual oven-proof casserole. Top with layers of potatoes, tomatoes, and onion; season to taste. Pour milk over each. Bake at 400° until milk bubbles. Top each casserole with 2 or 3 small drop biscuits. Bake at 400° for 15 minutes until biscuits are done. Makes 4 individual casseroles.

QUICK HAM LUNCH

PRESSED HAM LUNCH MEAT   1 5-ounce package, sliced

MARGARINE   2 tablespoons

CATSUP or TOMATO SAUCE   1/3 cup

WATER   1/3 cup

GREEN TOMATOES   1 cup chopped

GREEN PEPPER   1/2, sliced thin

COOKED NOODLES

BUTTER

PARMESAN CHEESE   grated

Cut ham in little squares. Saute in margarine until lightly browned. Add catsup, water, tomatoes, and green pepper. Simmer slowly until vegetables are tender. Serve on hot buttered noodles; sprinkle with Parmesan cheese. Serves 2.

MEAT LOAF SQUARES

ONION   1 cup chopped
GREEN BEANS   1/2 cup cut, raw or cooked
CELERY   1/2 cup sliced thin
VEGETABLE OIL   2 tablespoons
GREEN TOMATOES   1 cup chopped
LEAN GROUND BEEF   1 pound
PORK SAUSAGE   1/2 pound
WHEAT GERM   1/2 cup
SOFT BREAD CRUMBS   1/2 cup
SALT   1 teaspoon
FRESH GROUND PEPPER   1/2 teaspoon
GARLIC POWDER   1/2 teaspoon
WHOLE EGG   1
WORCESTERSHIRE SAUCE   1 teaspoon
TOMATO SAUCE   1/4 cup

Saute onion, beans, and celery in oil for 5 minutes. Add tomatoes and saute 5 minutes more. Cool. Mix rest of ingredients together; add vegetables. Pat into shallow 7 x 12-inch baking pan. Bake at 400° for 30 to 35 minutes, until nicely browned. Serves 8.

MEAT-LINED PIES

Crust

LEAN GROUND BEEF   1 pound
QUICK COOKING OATS   1/2 cup
TOMATO or BARBECUE SAUCE   2/3 cup
ONION   1, chopped fine
OREGANO   1-1/2 teaspoons
SEASONED SALT   1 teaspoon

Combine all ingredients; mix well. Press evenly into 2 oiled 9-inch pie pans on the bottom and up the sides.

Filling

MEDIUM-SIZE GREEN TOMATOES   3, sliced very thin
MARGARINE   3 tablespoons
WHOLE EGGS   4, beaten
MOZZARELLA CHEESE   1-1/2 cups shredded (or other
good melting cheese)
SALT   1 teaspoon
POTATOES   2 cups chopped and cooked

Saute tomatoes lightly in margarine. Layer tomatoes over the crusts. Mix eggs, cheese, salt, and potatoes. Spoon over the tomatoes, dividing evenly between the 2 pies. Bake at 375° until filling is set and meat crust is done. Serves 8.

STUFFED TOMATOES

LARGE GREEN TOMATOES   4
LEAN GROUND BEEF   1/4 pound
MEDIUM GREEN PEPPER   1/4, chopped
ONION   1 tablespoon chopped
SEASONED SALT   1/2 teaspoon
CHILI POWDER   1 teaspoon
WHOLE EGG   1, beaten
SOFT BREAD CRUMBS   4 tablespoons
MARGARINE   4 teaspoons

Slice off tops of tomatoes; remove and discard seeds and pulp, leaving a shell for stuffing. In a hot skillet, saute beef a few minutes. Drain off excess fat. Add green pepper, onion, and seasonings. Cook, stirring until beef is browned. Remove from heat; stir in egg. Spoon into tomato shells. Top with crumbs and margarine. Bake at 350° for 30 minutes, or until tomatoes are tender. Serves 2 to 4.

HAMBURGER CURRY

> ONION   1, chopped
> GROUND BEEF   1 pound
> VEGETABLE OIL   1 tablespoon
> CURRY POWDER   2 teaspoons
> GARLIC CLOVE   1, minced
> TOMATO SAUCE   1 8-ounce can
> WATER   1 cup
> GREEN TOMATOES   2-1/2 cups chopped
> SALT and PEPPER   to taste
> SUNFLOWER SEEDS   1/2 cup
> HOT RICE
> GREEN TOMATO CHUTNEY (page 74)

Saute onion and beef in oil, stirring, until lightly browned. Drain off excess fat. Add curry powder, garlic, tomato sauce, water, tomatoes, and seasonings. Simmer 15 to 20 minutes until tomatoes are tender. Add sunflower seeds. Leave mixture at room temperature 1/2 hour or more to blend flavors. Reheat and serve on hot rice with Green Tomato Chutney. Serves 4.

FILLED PANCAKE SUPPER

LEAN GROUND BEEF   3/4 pound
SMALL ONION   1, chopped
VEGETABLE OIL
MEDIUM-SIZE GREEN TOMATOES   6, cut in thin
wedges

RIPE OLIVES   1/2 cup sliced
SALT and PEPPER
DAIRY SOUR CREAM   1/2 cup
SHARP CHEESE   1 cup shredded
THIN PANCAKES

Brown beef and onions in oil; pour off excess fat. Add tomatoes and olives.
Saute until tomatoes are tender. Season to taste. Add sour cream.

Make thin pancakes in 7- or 9-inch skillet. Spoon filling in center of each
pancake, roll up, and place, seam-side down, in large oiled baking pan. Bake
at 350° for 25 minutes. Sprinkle cheese across pancakes and bake 5 minutes
more. Serves 4.

TOP-OF-THE-STOVE MEAT LOAF

LEAN GROUND BEEF   1-1/2 pounds
QUICK COOKING OATS   1-1/4 cups
GREEN TOMATOES   1 cup ground and drained (reserve
juice)
SALT   1 teaspoon
GARLIC POWDER   1/4 teaspoon
FRESH GROUND PEPPER   1/4 teaspoon
SMALL ONION   1, chopped fine
PARSLEY   1 tablespoon chopped
WHOLE EGG   1
GREEN TOMATO JUICE plus TOMATO JUICE   to make
1 cup
VEGETABLE OIL   2 tablespoons
ONION RINGS   sauteed

Combine all ingredients except oil and onion rings. Shape into a large rounded patty, slightly smaller than large skillet. Pour oil into hot skillet; place meat patty in hot oil, and cover tightly. Cook over medium heat for 15 minutes. Loosen with spatula and turn patty over, browned side up. Cut into 6 wedges, separating them slightly. Cover and cook 15 minutes, or until done. Serve topped with onion rings. Serves 6.

NOODLE LUNCH

BACON   4 slices, chopped
ONION   1, sliced thin
LARGE GREEN TOMATOES   4, chopped
GREEN PEPPER   1, cut in strips
PRESSED HAM LUNCH MEAT   1 5-ounce package, slivered

RIPE OLIVES   1 cup sliced
TOMATO JUICE   1 15-ounce can
NOODLES   1 6-ounce package
SHARP CHEDDAR CHEESE   1/4 cup shredded

Saute bacon until browned; remove and keep warm. Fry onion in bacon fat until golden; pour off half the fat. Add tomatoes, green pepper, ham, olives, and tomato juice. Bring to boil; add noodles. Cover and simmer until noodles are tender. Stir in cheese and bacon. Serve at once. Serves 4.

MRS. PITRE'S SMOTHERED GREEN TOMATOES

PORK STEAKS   4
VEGETABLE OIL or BACON DRIPPINGS   2 tablespoons
MEDIUM-SIZE GREEN TOMATOES   6, diced
WATER   1/2 cup
CHICKEN BOUILLON CUBES   2
SALT and PEPPER   to taste
SUGAR   2 teaspoons
POWDERED FENUGREEK   1/2 teaspoon
RICE   cooked

Brown steaks on both sides in oil in large skillet. Add tomatoes, water, bouillon, and seasonings. Cover and let simmer until steaks are very tender. Serve with rice. Serves 4.

CHICKEN IN SAUCE

BACON   *4 slices, chopped*
FRYER   *1, cut up (or 8 CHICKEN THIGHS)*
FLOUR   *2 tablespoons*
POULTRY SEASONING   *1/2 teaspoon*
THYME   *1/2 teaspoon crushed*
GARLIC POWDER   *1/2 teaspoon*
SALT   *1/2 teaspoon*
ONION   *1, sliced thin*
CONDENSED TOMATO SOUP   *1 10-1/2-ounce can*
CHICKEN BOUILLON CUBES   *2, crushed*
LARGE GREEN TOMATOES   *6, chopped (or 12 small ones)*
FRESH GROUND PEPPER   *sprinkle*
GINGER   *pinch*
SUGAR   *pinch*

Brown bacon in large skillet. Remove bacon and reserve. Shake chicken in bag with flour and seasonings; then saute it in bacon drippings in skillet until golden. Add bacon, onion slices separated into rings; saute until onion is limp. Add tomato soup, bouillon cubes, tomatoes, pepper, ginger, and sugar. Cover and simmer slowly until chicken is very tender, stirring occasionally. Serve chicken with its thick sauce. Serves 4.

CHICKEN AND RICE

FRYER   1, cut in pieces (or 8 CHICKEN THIGHS)
VEGETABLE OIL   3 tablespoons
SALT
PAPRIKA
ONIONS   2, chopped
WATER   3 cups
CHICKEN BOUILLON CUBES   2
RICE   1 cup uncooked
LARGE RIPE TOMATOES   2, sliced
LARGE GREEN TOMATOES   6, chopped (or 12 small
ones)
FROZEN PEAS   1 10-ounce package
SAFFRON   1/4 teaspoon
CURRY POWDER   1/4 teaspoon

In Dutch oven brown chicken in oil. Pour off excess oil. Sprinkle chicken
with salt and paprika. Add onion; fry, stirring, until onion is limp. Add water
and bouillon cubes; bring to boil. Stir in rice, tomatoes, and peas; add
seasonings. Bring to boil again. Cover tightly and bake at 350° for 1/2 hour or
more, until rice is tender. Stir gently from time to time. Serves 4.

SPANISH CHICKEN STEW

LARGE FRYER or STEWING HEN   1
SALT and PEPPER
PAPRIKA
LARGE SPANISH ONION   1, chopped
VEGETABLE OIL   3 tablespoons
CHICKEN BOUILLON CUBE   1
HOT WATER or TOMATO JUICE   1/4 cup
LARGE RIPE TOMATOES   2, chopped
LARGE GREEN TOMATOES   3, chopped
GREEN PEPPER   1, sliced
CARROTS   1 cup, chopped
CELERY STALKS   3, sliced
MUSHROOMS   1/2 pound, sliced
FROZEN PEAS   1 10-ounce package, thawed
RIPE OLIVES   1/2 cup sliced
PARSLEY   garnish
HOT SAFFRON RICE

Cut chicken into serving pieces; season with salt, pepper, and lots of paprika. Saute onion in oil in large Dutch oven. Add chicken and brown lightly. Pour off excess fat. Add bouillon cube, dissolved in hot water. Cover and cook on low heat for 1/2 hour (or longer if stewing hen). Add tomatoes, green pepper, carrots, and celery. Cover and cook slowly until all are tender. Add mushrooms, peas, and olives. Cover and cook 15 minutes more. Season to taste with salt and pepper. Serve with hot saffron rice. Garnish with parsley. Serves 6.

PICKLES

RELISHES

DILLED GREEN TOMATO PICKLES

TINY GREEN TOMATOES or GREEN CHERRY
TOMATOES

Wash tomatoes, prick with a fork or needle, and pack into hot washed pint jars.

Add to each jar:

> *GARLIC CLOVE   1*
> *FRESH DILL HEAD   1 (or DILL WEED 1 teaspoon)*
> *HORSERADISH ROOT   1 slice*

Combine and bring to boil:

> *WATER   1 quart*
> *VINEGAR   2 quarts*
> *PICKLING SALT   3/4 cup*
> *ALUM   1/2 teaspoon*

(This amount of liquid fills 10 to 12 pint jars.)

Pour hot liquid over tomatoes, filling to within 1/2 inch of top of jar. Seal. Process in boiling water bath 10 minutes. Ready to use in 6 weeks.

END-OF-THE-GARDEN PICKLES

GREEN TOMATOES   8 quarts chopped
PICKLING SALT   1 cup
CELERY   2 quarts sliced
CABBAGE   4 quarts chopped
CAULIFLOWER FLOWERLETS   2 quarts
CUCUMBERS   4 quarts sliced or in chunks
ONIONS   2 quarts sliced
GREEN PEPPERS   2 quarts cut in strips
DARK BROWN SUGAR   2 cups, firmly packed
POWDERED HORSERADISH   1 teaspoon
CINNAMON   1 teaspoon
ALLSPICE   1 teaspoon
CLOVES   1 teaspoon
MACE   1 teaspoon
GINGER   1/2 teaspoon
MUSTARD SEED   3 tablespoons
CELERY SEED   1 tablespoon
DRY MUSTARD   3 tablespoons
CIDER VINEGAR   3 to 4 quarts (enough to cover vegetables)

In large crock or enamel pan, sprinkle tomatoes with salt, and let stand 8 hours or overnight. Drain well. Add other vegetables; mix in sugar, seasonings, and vinegar. Bring to a boil and boil 5 minutes, stirring gently. Ladle into washed pint jars; seal. Process 15 minutes in boiling water bath. Ready to use in 6 weeks.

GREEN TOMATO TACO SAUCE

GREEN TOMATOES   8 quarts chopped
CAYENNE PEPPER   1 teaspoon
CIDER VINEGAR   2 quarts
SALT   4 tablespoons
FRESH GROUND PEPPER   4 tablespoons
DRY MUSTARD   1 tablespoon
CLOVES   1 tablespoon
ALLSPICE   1 tablespoon

Cook tomatoes and cayenne in vinegar until tomatoes are soft. Then put through food mill. Add seasonings and boil slowly 5 hours, stirring frequently to prevent sticking. When thick, pour into washed 1/2-pint jars. Seal and process in boiling water bath for 5 minutes.

Note: This sauce, similar to Salsa Verde, is excellent on hot dogs and hamburgers, with fresh shredded vegetables, in salad dressings, deviled eggs, heuvos rancheros, etc.

Recipe from   BAINBRIDGE ARTS AND CRAFTS STARVING ARTISTS COOK BOOK, Bainbridge Arts and Crafts, Inc., Bainbridge Island, Washington, 1975.

GREEN TOMATO CHILI SAUCE

GREEN and RIPE TOMATOES   10 cups (about 6 cups
                green and 4 cups ripe, whatever you have on hand)
GREEN PEPPERS   2, chopped
ONIONS   5, chopped
GARLIC CLOVES   2, mashed
VINEGAR   2 cups
CINNAMON   2 tablespoons
CLOVES   2 teaspoons
GINGER   2 teaspoons
PAPRIKA   2 teaspoons
DRY MUSTARD   1 teaspoon
CURRY POWDER   1/2 teaspoon
SUGAR   3 tablespoons
SALT   1 tablespoon (optional)

Chop vegetables or grind them in a food grinder. Combine all ingredients in enamel or stainless steel pan. Simmer on low heat for 1 hour. Then put through food mill or sieve. Put some of the pureed mixture in the blender and add the pulp left in the food mill. Hold blender top down securely with a towel so hot liquid does not force the top off. Blend; then add to rest of puree. Simmer for 1-1/2 to 2 hours more until thick. Pour into hot washed jars and seal; or use hot washed soft drink bottles, sealing with bottle capper and crown caps dipped in boiling water. Makes about 7 cups.

Note: This is a good condiment for a low-sodium diet, if you omit the salt or use about 1/2 teaspoon low-sodium salt substitute instead of salt.

MARTHA'S VINEYARD OLD-TIME MINCEMEAT*

GREEN TOMATOES   8 quarts coarsely ground or chopped fine

COLD WATER

VINEGAR   3/4 cup

FROZEN CONCENTRATED APPLE JUICE   1/4 cup

CINNAMON   2 tablespoons

CLOVES   2 tablespoons

ALLSPICE   2 tablespoons

NUTMEG   1 tablespoon

CANDIED CITRON   1/2 pound, chopped

Drain tomatoes, and replace juice with cold water. Let come to a boil; drain off. Repeat 2 or 3 times, adding the same amount of cold water each time, until tomatoes are clear. Then add vinegar, apple juice, and spices. Cook slowly about 2 hours. Add citron. Cook slowly 2 hours more, or until tomatoes are about the color of raisins. Makes about 5 quarts of mincemeat. Preserve by freezing or canning.

Note: When using this mincemeat in some of the dessert recipes, you may want to add a little honey or brown sugar to recipe, since this is not very sweet.

*Prize recipe of Gertrude Turner's mother.

GREEN TOMATO MINCEMEAT

GREEN TOMATOES   8 cups coarsely ground and undrained

FROZEN CONCENTRATED APPLE JUICE or TOMATO
JUICE

APPLES   8 cups coarsely ground, cored but unpeeled

RAISINS   2 pounds

DATES   1 pound, chopped

BROWN SUGAR   2 cups, firmly packed

VINEGAR   1-1/3 cups

CINNAMON   1 tablespoon

ALLSPICE   3/4 teaspoon

CLOVES   1 teaspoon

MACE   3/4 teaspoon

PEPPER   1/4 teaspoon

SALT   1 teaspoon (optional)

VEGETABLE OIL   2/3 cup

Drain tomatoes, and replace juice with apple juice or tomato juice, adding a
little more sugar for the latter. Mix all ingredients together, except oil.
Simmer until thick and flavors are blended. Add oil and stir well. Spoon into
freezer containers, or into hot washed pint jars and process 25 minutes in
boiling water bath.

GREEN TOMATO CHUTNEY

GREEN TOMATOES   10 cups chopped
APPLES   4 cups chopped or ground, unpeeled
VINEGAR   4 cups
BROWN SUGAR   2 pounds (4-2/3 cups, firmly packed)
SALT   1 tablespoon
LARGE ONIONS   2, chopped
GARLIC CLOVE   1, minced or crushed in press
GINGER   1 teaspoon
ALLSPICE   1 teaspoon
MUSTARD SEED   2 teaspoons
PEPPER   1/2 teaspoon

Combine ingredients. Heat slowly until sugar is dissolved. Simmer until mixture is thick and clear. Pour into hot washed jars and seal at once.

MUSTARD-PICKLE RELISH

GREEN TOMATOES 3 quarts chopped and drained
CUCUMBERS 1 quart sliced or cubed
LARGE CAULIFLOWER 1, cut in pieces
CELERY 1 bunch, cut in 1-inch slices
ONION 2 cups chopped
PICKLING SALT 1 cup

Add salt to vegetables and let stand overnight; then drain. Cover with fresh water; bring to a boil; boil about 5 minutes and drain.

Mustard Sauce

DRY MUSTARD 1/2 cup
FLOUR 1/2 cup
TURMERIC POWDER 1 tablespoon
SUGAR 3/4 cup
CIDER VINEGAR 3 pints

Combine dry ingredients. Add vinegar slowly, stirring until smooth. Cook slowly for 1/2 hour, stirring to keep from sticking. Pour this over drained vegetables; bring to a boil. Pack into hot washed pint jars and seal. Process 20 minutes in boiling water bath.

Note: String beans, cabbage, or other garden vegetables may be added.

GREEN TOMATO PORK
OR CHICKEN RELISH

GREEN TOMATOES   4 cups quartered
FROZEN CONCENTRATED APPLE JUICE   2/3 cup
ORANGE   1, cut in pieces
RAISINS   1/2 cup
CINNAMON   1 teaspoon
CLOVES   1/2 teaspoon

Place tomatoes in blender container with apple juice; blend. Add orange pieces, and chop by turning blender on and off a few times. Turn into saucepan; add raisins and spices. Simmer 15 to 30 minutes, or until thick. Ladle into hot washed pint jars and seal. Process in boiling water bath for 20 minutes.

No sugar or salt in this recipe. Great for dieters.

HOT BURGER RELISH

CHILI POWDER   3 tablespoons
DRY MUSTARD   3 teaspoons
SALT   1 tablespoon
VINEGAR   3 cups
GREEN TOMATOES   3 quarts chopped
LARGE GREEN PEPPERS   3, chopped
LARGE ONIONS   2, chopped
HORSERADISH   1 tablespoon

Boil up chili powder, mustard, salt, and vinegar. Add rest of ingredients; boil 5 minutes. Pack into hot washed pint jars and seal. Keeps well in refrigerator, but process what you will not use up in a month. Process pints 10 minutes in boiling water bath.

MRS. ARMBRUSTER'S
GERMAN CHOWDER RELISH

GREEN TOMATOES   1 gallon chopped fine
SALT   2/3 cup
MEDIUM ONIONS   8, chopped
LARGE GREEN PEPPERS   4, chopped (or 2 GREEN
                    PEPPERS and 2 RED SWEET PEPPERS)
SUGAR   4 cups
CIDER VINEGAR   2-2/3 cups
CLOVES   1 teaspoon
CINNAMON   2 teaspoons
CELERY SEED   2 teaspoons

Sprinkle tomatoes with salt; let stand overnight in glass or crockery bowl. Then drain; add rest of ingredients. Bring to a boil, and boil for a few minutes. Ladle into washed pint jars. Process for 10 minutes in boiling water bath.

GREEN TOMATO RAG RELISH

GREEN TOMATOES   3 quarts cut fine
CABBAGE   3 quarts chopped fine
ONIONS   6, chopped
CELERY SEED   1 tablespoon
DRY MUSTARD   1 tablespoon
TURMERIC POWDER   1 tablespoon
SALT   1 tablespoon
SUGAR   3 cups
VINEGAR   6 cups

Combine all ingredients. Simmer 1/2 hour, stirring from time to time. Pour into hot washed pint jars and seal. Process 10 minutes in boiling water bath.

GREEN TOMATO JAM

SUGAR   6 cups
GREEN TOMATOES   3 pounds, chopped (stem end re-
moved)
LEMONS   2, juice and grated peel
CINNAMON STICKS   2

Warm sugar in a pan in oven for a few minutes, and then add to tomatoes. Mix and let stand 1 hour to draw out the juice. Add lemon juice and peel, and cinnamon sticks. Boil, stirring often, until thick and transparent. Remove cinnamon sticks. Pour into hot washed jars and seal at once. Makes 5 half-pint jars.

GREEN TOMATO MARMALADE

LEMON   1

WATER

GREEN TOMATOES   1 quart sliced very thin (stem end removed)

SUGAR   1-1/4 cups

FROZEN CONCENTRATED APPLE JUICE   1 tablespoon

CANDIED or SUGAR-GINGER   1 piece, chopped (optional)

Peel lemon with vegetable peeler and reserve peel. Slice lemon thin. Cut lemon peel into strips and boil in 1 cup water. Discard water and repeat several times adding the same amount of water each time; drain. Mix tomatoes, sugar, and apple juice. Add the lemon peel and slices. Heat slowly, stirring often, until thick and transparent, about 1 to 1-1/2 hours. Add sugar-ginger and simmer a few minutes more. Remove from heat and let stand overnight. Then reheat to boiling; pour into hot washed jars and seal at once.

VEGETABLES

SIDE DISHES

SPINACH-STUFFED TOMATOES

LARGE GREEN TOMATOES   4
SPINACH   1 cup cooked, chopped, and drained
SEASONED SALT   1/4 teaspoon
ONION POWDER   1/4 teaspoon
DAIRY SOUR CREAM   2 tablespoons
CATSUP   2 tablespoons
PARMESAN CHEESE   grated

Slice off tops of tomatoes. Scoop out seeds and pulp and discard, leaving a shell for stuffing. Combine spinach, salt, onion powder, sour cream, and catsup. Stuff tomato shells. Bake at 375° for 30 minutes until tomatoes are tender. Sprinkle with cheese. Serves 2 to 4.

BROILED GREEN TOMATOES

MEDIUM-SIZE GREEN TOMATOES   1 per person as
    meat garnish; 2 per person as main vegetable
MAYONNAISE
THYME
DILL WEED
SEASONED SALT
PARMESAN CHEESE   grated

Cut tomatoes in half. Place, cut side up, on broiler rack. Spread with mayonnaise; sprinkle with thyme and dill weed, then with salt. Broil 3 to 4 inches from broiler element for about 15 minutes, or until tender. Sprinkle with cheese.

SCALLOPED CABBAGE AND TOMATOES

LARGE GREEN TOMATOES 3, sliced very thin
BROWN SUGAR 3 tablespoons
CABBAGE 3 cups shredded
VINEGAR 1 tablespoon
MARGARINE 4 tablespoons, melted
SALT 1/2 teaspoon
FRESH GROUND PEPPER 1/4 teaspoon
SOFT BREAD CRUMBS 1 cup, buttered

Layer alternately in greased casserole: 1 sliced tomato sprinkled with 1 tablespoon brown sugar; 1 cup cabbage sprinkled with 1/3 of vinegar and margarine mixture; and 1/3 cup crumbs. Salt and pepper each layer lightly. Repeat layers, finishing with layer of crumbs. Bake at 375° for 25 to 30 minutes, or until tender. Serves 4.

FRIED GREEN TOMATOES

LARGE GREEN TOMATOES   3
FLOUR   3 tablespoons
SEASONED SALT   1/2 teaspoon
MARGARINE or BACON FAT   3 tablespoons
SUGAR   1/2 teaspoon (optional)

Cut tomatoes into 1/3-inch slices. Dip both sides into flour and salt mixture. Saute in margarine until brown; turn carefully and brown other side. Sprinkle with sugar.

CURRIED GREEN TOMATOES

LARGE GREEN TOMATOES   3, cut in wedges
SMALL APPLE   1, cored, peeled, and chopped
ONION   1 large slice, chopped
MARGARINE   1 tablespoon
CHICKEN BOUILLON CUBE   1
HOT WATER   1 tablespoon
CURRY POWDER   1 teaspoon
SALT and PEPPER   to taste
COOKED RICE   1/2 to 1 cup (optional)

Saute tomatoes, apple, and onion in margarine for 5 minutes. Dissolve bouillon cube in hot water. Add with rest of ingredients to tomato mixture and cook slowly for a few minutes more, until tomatoes are tender. Serves 2.

FRENCH-FRIED GREEN TOMATOES

LARGE GREEN TOMATOES   3
FLOUR   3 tablespoons
SEASONED SALT   3/4 teaspoon
WHOLE EGG   1
WATER   2 tablespoons
DRY BREAD CRUMBS   1/2 cup
OIL   for deep fat frying

Slice tomatoes 1/2-inch thick. Dip in flour and salt mixture, then in egg beaten with water, and then in crumbs. Let dry a few minutes. Fry in deep fat until brown.

EASY CURRY SIDE DISH

MEDIUM-SIZE GREEN TOMATOES   10 (can use some
                                   turning pale pink)

MARGARINE   3 tablespoons
CURRY POWDER   2 teaspoons
SALT   to taste
FENUGREEK   pinch (optional)

Remove and discard stem ends of tomatoes. Do not peel. Cut each tomato into 6 or 8 wedges. Melt margarine in large skillet; add tomatoes. Sprinkle with curry powder, salt, and fenugreek. Stir to blend in seasonings and coat tomatoes with margarine. Saute 3 to 4 minutes, just until tomatoes are bubbly hot. Do not overcook. Serve at once. Serves 4.

GREEN TOMATOES WITH HERBS

CELERY   1/2 cup thinly sliced

MARGARINE   3 tablespoons

MEDIUM-SIZE GREEN TOMATOES   4 (or 8 small ones)

SMALL RIPE TOMATO   1

GARLIC CLOVE   1, crushed through press

PARSLEY   1/4 cup chopped

BASIL   1/4 teaspoon

THYME   1/4 teaspoon

SEASONED SALT   1/4 teaspoon

SESAME SEEDS   1 teaspoon

Saute celery lightly in margarine for 2 minutes. Remove and discard stems of tomatoes and cut tomatoes into wedges. Add tomatoes and rest of ingredients to celery. Saute until tomatoes are just tender. Do not overcook. Serves 4.

BAKED TOMATOES

MEDIUM-SIZE GREEN TOMATOES   2 per person
SHARP CHEDDAR CHEESE   cut in wedges, for stuffing
SOFT BREAD CRUMBS   buttered
SEASONED SALT

Remove tomato stem; hollow out center enough to insert little wedge of cheese. Cover cheese with crumbs; season. Bake in covered baking dish, with a little water in the bottom, at 375° for 40 to 45 minutes. Then brown tops under broiler, if desired.

ITALIAN FRIED TOMATOES

LARGE GREEN TOMATOES   3
PARMESAN CHEESE   1/4 cup grated
CORNMEAL   1/4 cup
FLOUR   1 tablespoon
SEASONED SALT   1 teaspoon
OREGANO   1 teaspoon
OLIVE OIL or VEGETABLE OIL
GARLIC CLOVE   1, minced into oil

Cut thin slice off top and bottom of each tomato and discard. Slice tomatoes into 1/2-inch slices. Dip into mixture of cheese, cornmeal, flour, and seasonings. Fry on both sides in oil, with garlic added, until nicely browned. Keep hot in oven until ready to serve; 350° if used within 10 minutes, 250° if used in 15 minutes.

SCALLOPED GREEN TOMATOES

MEDIUM-SIZE GREEN TOMATOES   6 (or 12 small ones)
ONION   1 slice, chopped fine
RIPE TOMATO   1, chopped
SALT and PEPPER   to taste
SUGAR   to taste
DRY BREAD CRUMBS   1 cup
MARGARINE   1 tablespoon

Cut green tomatoes in wedges. Put in saucepan with onion, ripe tomato, and seasonings. Cover tightly; cook VERY slowly for about 5 minutes, shaking pan occasionally. Place half of crumbs in shallow oiled baking dish. Spoon tomato mixture over crumbs. Mix rest of crumbs with margarine, and sprinkle over tomato mixture. Bake at 400° for about 15 minutes, or until nicely browned. Serves 4.

VEGETABLE CASSEROLE

LARGE GREEN TOMATOES   4, thinly sliced
MEDIUM ONIONS   2, thinly sliced
SEASONED SALT
MARGARINE   2 tablespoons
EVAPORATED MILK   1/2 cup
PAPRIKA

Alternate layers of tomato and onion in shallow oiled casserole. Sprinkle with salt and dot with margarine. Cover and bake at 350° for 15 minutes. Pour milk over it, and sprinkle liberally with paprika. Bake uncovered until brown and tender.

88

DEVILED GREEN TOMATOES

MEDIUM-SIZE GREEN TOMATOES   3
DRY MUSTARD   1 teaspoon
VINEGAR   1/2 teaspoon
POWDERED SUGAR   1/2 teaspoon
SEASONED SALT   1/2 teaspoon
MAYONNAISE   3 tablespoons
PARMESAN CHEESE   1 teaspoon grated

Cut thin slice off top and bottom of each tomato and discard. Cut each tomato into 3 thick slices. Place on broiler rack. Mix rest of ingredients together. Brush tomatoes lightly with the mixture. Broil several inches from broiler element for 3 minutes. Turn carefully. Spread tomatoes with thick layer of mixture and broil. When topping has bubbled into tomatoes for a minute or so, spread again with remaining mixture. Broil until browned and tender.

SOUTHERN FRIED TOMATOES

LARGE GREEN TOMATOES   3
WHOLE EGG   1, beaten
EVAPORATED MILK   2 tablespoons
FINE DRY BREAD CRUMBS   1/3 cup
YELLOW CORNMEAL   1/3 cup
BRAN   1/3 cup (not boxed cereal)
SALT   1/4 teaspoon
SUGAR   1/4 teaspoon
MARGARINE or BUTTER   3 tablespoons
FRESH GROUND PEPPER

Cut thin slice off top and bottom of each tomato and discard. Slice tomatoes into 1/3-inch slices. Beat egg with milk. Combine crumbs, cornmeal, bran, salt, and sugar. Dip tomato slices in egg mixture, then dip in dry ingredients. Fry in margarine in large skillet, turning to brown both sides. Sprinkle heavily with pepper. Serve hot.

SEPTEMBER STIR-FRY

VEGETABLE OIL   3 tablespoons
CELERY STALKS   4, diagonally sliced
FROZEN ASPARAGUS SPEARS   6 large, diagonally
                                       sliced thin
MEDIUM-SIZE GREEN TOMATOES   4, cut in wedges
MUSHROOMS   1/4 pound, sliced
MEDIUM JERUSALEM ARTICHOKES   4, peeled,
                                       quartered, thinly sliced
FRESH BEAN SPROUTS   1 cup
SUGAR   1/2 teaspoon
SOY SAUCE   3 tablespoons

Heat wok or skillet. Add oil and stir-fry celery and asparagus for 3 minutes, stirring constantly. Add tomatoes, mushrooms, artichokes, and bean sprouts. Saute 2 minutes, stirring constantly. Add sugar and soy sauce; toss lightly. Serve at once. Vegetables should be tender-crisp. Serves 4.

Note: This could be a luncheon main dish with addition of cooked shrimp, julienne slices of ham, or other cooked meat.

INDEX

MEAT

COOKIES

DESSERTS

Paula Simmons is a handspinner and weaver. For over 19 years, she and her husband have raised black sheep on their farm near Suquamish, Washington, and spun the wool into yarn for sale to knitters and weavers. Wasted hay and barn cleanings were spread on the garden, producing a seasonal abundance that demanded action. Mrs. Simmons is a devoted recipe-creator and recipe-swapper; cookbooks soon blossomed along with the produce.

Ruth Richardson, a Northwest artist, does silk-screening of calendars, book covers, gift wrap paper, and serigraphs. In her spare time she raises goats and gardens.